PUFFIN BOOKS
NO HOLIDAY FUN FOR SAM

Sam and his aunt and uncle are going on holiday to the seaside. Sam has been looking forward to this holiday for ages, but his heart sinks when he sees the words NO BUCKETS AND SPADES IN THE HOUSE in their hotel. It looks like he might be in for a dismal holiday, but will it turn out all right for Sam? Then, on top of that, Sam's cub pack have planned a camping trip to Wales. Sam is understandably thrilled – for weeks he thinks of nothing else: his big rucksack is packed and ready in the hall, surely nothing can go wrong now to spoil his fun?

These two enjoyable stories about the unstoppable Sam will entertain and amuse young readers.

Thelma Lambert is an artist and designer who lives in London with her husband and family.

No Holiday Fun
for Sam

Written and illustrated by
THELMA LAMBERT

PUFFIN BOOKS

PUFFIN BOOKS

Published by the Penguin Group
Penguin Books Ltd, 27 Wrights Lane, London w8 5tz, England
Viking Penguin, a division of Penguin Books USA Inc.
375 Hudson Street, New York, New York 10014, USA
Penguin Books Australia Ltd, Ringwood, Victoria, Australia
Penguin Books Canada Ltd, 2801 John Street, Markham, Ontario, Canada l3r 1b4
Penguin Books (NZ) Ltd, 182–190 Wairau Road, Auckland 10, New Zealand

Penguin Books Ltd, Registered Offices: Harmondsworth, Middlesex, England

No Train for Sam first published by Hamish Hamilton Children's Books 1989
No Camping for Sam first published by Hamish Hamilton Children's Books 1988
Published in one volume as *No Holiday Fun for Sam* in Puffin Books 1991
1 3 5 7 9 10 8 6 4 2

Filmset in Baskerville
Printed in England by Clays Ltd, St Ives plc

Contents

NO TRAIN FOR SAM

For Annie-Annie

Chapter One

"Uncle Den! I feel sick!" moaned
Sam from the back of the car.

"Nearly there now, old chap!"
said Uncle Dennis, putting on a
burst of speed.

Eight-year-old Sam and his aunt
and uncle were on their way to
Devon for the whole summer, but
the little car Uncle Dennis had

borrowed was a bit of a bone-shaker. The journey seemed to be going on forever!

"Fifty pence for the first person to see the sea," said Aunty Kathleen.

Sam had been looking forward to this holiday for ages. He lived in a flat in the city, so going to the sea-side was a big treat. They were going to stay at a small hotel called 'Seaview'. Sam thought it would be a cottage with roses round the door, right next to the sea. Just then Sam saw a glimpse of blue.

"AUNTY! I CAN SEE THE SEA!" he cried. "Fifty pence, please!"

"Feeling better, Sam?" smiled Uncle Dennis.

It was raining when they finally
arrived. They found 'Seaview' down
a back street. It was an ugly, red-
brick villa. 'Bed and Breakfast', said
a sign in the window.

"Do they only have breakfast
here?" said Sam in dismay. "What
about dinner, and tea?"

"We'll find a fish and chip shop
when we've unpacked," said his
aunt.

A woman opened the door. She was Mrs Buller, the landlady. And the first thing Sam saw was a notice: *No Buckets and Spades in the House*. Sam didn't feel car-sick anymore; he felt *home*-sick . . .

They went upstairs to unpack.

No Buckets and Spades in the House

"Uncle! You can't see the sea from any of the windows," said Sam, falling off a chest of drawers.

"Never mind, you can go to the beach every day while we're here," said his uncle.

It was getting dark already and the rain was coming down harder than ever as they trudged down the High Street. The fish and chip shop was closed.

"Never mind," said Aunty Kathleen. "I've still got lots of sandwiches back at the house."

"There's sure to be a nice breakfast," said Uncle Dennis, trying to be cheerful.

They went down to look at the sea. The tide was in and big dark waves lashed the sea-front. There was no-one about except a stray dog, who looked very wet and sorry for itself. Soaked through, they trailed miserably back to 'Seaview'.

Chapter Two

Sam woke to the sound of rain
against his window. He groaned but
the thought of breakfast cheered him
up. He got dressed and raced
downstairs only to find Mrs Buller
was serving kippers for breakfast.

"Don't like kippers," muttered
Sam.

When no one was looking he hid

the kipper in his pocket. Later he
fed it to Mrs Buller's cat, who was
promptly sick on the carpet.

"I can see I'll have to keep my
eye on *you*," said the landlady to
Sam, as she mopped up.

Sam took his bucket and spade
and went down to the sea. He had
been told not to go in the water, just
to play on the beach. But the tide
was coming in and there was only a

narrow strip of beach left. Sam sat
throwing pebbles into a rockpool,
and wondered what to do. Just then
he heard a sound which made him
jump up.

"Toot, toot," it went. "Toot
TOOT!"

A little red railway engine, puffing
out smoke like blobs of cotton wool,
was coming along the sea-front! The
carriages were full of children, who
waved to Sam, and he waved back.

He ran over to where the engine
was just pulling into a low platform.
As it stopped, it let out a great
"Ooouuufff!" of steam, as if well-
pleased with itself.

Sam jingled his ice-cream money
in his pocket; if he went without
ice-cream he would have enough for
a ride! He joined the queue of

excited children, bought a ticket and climbed aboard. A whistle blew and with a hiss of steam they were off.

At that moment the sun came out from behind a cloud. The little train went rattling along the edge of the seafront. Sam leaned out of the window and watched all the people waving at the train. They passed lots of shops and some big hotels. Then,

when they reached the pier, the
train did a loop and returned to its
starting point. Sam thought it was
the most exciting thing in the world.

After the ride he couldn't wait to tell his aunt and uncle all about the little train. In his hurry to get back to the hotel, he fell into a rockpool, and arrived back dripping wet. Mrs Buller opened the door and glared at his wet shoes.

"And what is *that*? TAKE IT OUT OF HERE AT ONCE!"

She pointed to Sam's bucket, which had toppled over. A small crab slid out and scuttled over the landlady's slippers. She gave a screech which brought Uncle Dennis on the scene. Excitedly Sam told him all about the steam train.

"Wonderful news, old chap!" said Uncle Dennis.

After that Sam didn't buy any more
ice-creams: instead he bought tickets
for the little train. Every morning he
would sit in the front carriage right
behind Ted, the engine-driver. Sam
wanted to learn all he could about
steam trains and how they worked.

One day the lad who usually
helped Ted with the train didn't
turn up. Ted saw Sam standing as

he always did, watching everything.

"Hey you! Sam, isn't it? How would you like to be my guard today?"

Sam didn't need to be asked twice! He clambered aboard. Ted let him wave the green flag and blow the whistle, and the little red train pulled slowly out of the platform.

From then on Sam was Ted's regular helper. He would wave the green flag, clean the brass bits on the engine, and fetch the coal. On some days, when Sam had to break the coal into small lumps, he came back to 'Seaview' covered in coal-dust. Mrs Buller would make him wash in a bucket of cold water in the back yard before she allowed him in the house. But Sam didn't mind; it was worth it!

Ted and Sam became firm friends.
Ted got to know Uncle Dennis and
Aunty Kathleen too. They often
shared fish and chips at the café.

"Your Sam is a great little
helper," said Ted. "I reckon he
knows so much about this train he
could drive her himself!"

Then Sam and his aunt and uncle had to go away for a few days, to visit Great Uncle George who lived in Taunton. Sam didn't want to go: he couldn't bear to leave the little train.

"Cheer up, Sam!" said Aunty Kathleen. "It's not for long."

However it was more than a week before they returned. The old car had broken down and Uncle Dennis had had to take it to a garage. When they got back to 'Seaview' at last, Sam leapt out of the car. His aunt and uncle didn't need to ask where he was off to – they knew he would be going straight to the little train.

But when Sam got to the sea-front there was NOTHING THERE! No little train waited for passengers, puffing out blobs of smoke; no excited children clutched their tickets; no railway track ran along by the sea. The sign *Ted's Little Train* had been taken down and lay half buried in the sand.

He couldn't believe it. NO
TRAIN FOR SAM . . . ?

He walked slowly back to the
hotel. His aunt was in the hall,
talking to Mrs Buller. She knew all
about it! It was the talk of the town!
She told them Mr Crossman, who
lived in that big house on the hill,
had bought the train for his
children! He had had the railway
track laid all round his own garden.

Sam was devastated.

"How could Ted sell the train,
Aunty?" said Sam.

"Well, I expect he needed the
money," sighed Aunty Kathleen.

Chapter Three

Sam longed to see the little train
again. The chance came the very
next day when Uncle Dennis
suggested Sam might like to go and
fly his kite up Barton Hill. On the
way, Sam passed Barton Hall, the
big white house where Mr Crossman
lived. He came to a thick, high
hedge, and seeing a gap, he peered

in. Spread out before his gaze was a
velvety lawn as big as a cricket
pitch! There were beautifully kept
flower-beds and a pond where a
fountain played. All at once Sam
heard a familiar sound. "Toot toot,"
it went. "Toot TOOT!"

The little red engine came puffing
into view. Three children were

riding on the train and what a lot of
noise they were making, squabbling
and fighting! The smallest child got
pushed off into a rose bush and set
up a loud wailing. At that moment a
tall man came out of the house to
see what all the noise was about. It
was the children's father.

"I wish I'd never bought you that wretched train!" he shouted angrily. "All you do is quarrel over it!"

Sam squeezed back through the hedge before he could be seen. Carrying his kite, Sam hurried back to the beach and told his aunt and uncle all he'd seen.

A few days later Sam went up Barton Hill to fly his kite again. He found himself passing the high hedge of the Crossmans' garden once more. He couldn't resist peeping in again.

This time there were no noisy children quarrelling; just the bees buzzing in the flowers and the water splashing in the fountain. An old gardener was working his way slowly along a flower border.

"Hello!" Sam called to him.

The old man pretended not to hear. Sam tried again.

"HELLO! Excuse me, may I go and look at the train please? I used to help Ted look after it . . ." Sam's voice trailed away.

The gardener stopped his work.
He told Sam the family had all gone
away on holiday, and he was only
Crabtree: he didn't know anything
about the children's train. It was
nothing to do with *him*. He had
enough to do as it was . . .
Grumbling away to himself, he went
back to his digging.

"Mr Crabtree! Could I just have a
look? *Please*?"

"Suppose so," muttered the old
man. "Nothing to do with me if it's
all rusting away."

Sam ran over the lawn to where
the train stood. He pulled out a
hanky and began to wipe the engine
down.

"It does seem a shame," said Mr

Crabtree. "Could give 'er a rub with this," and, holding out a tin of Brasso, the old man ambled off.

Sam set to work.

After that Sam went back almost every day to look at the train. He liked to 'help' Mr Crabtree in the garden, too. Sam couldn't tell the difference between a plant and a weed; but he could learn, couldn't he? Sam became a real helper, and only fell in the pond twice.

"It's many years since I've had a

gardener's boy," said the old man, as they shared a lunch of bread and cheese.

One morning when Sam went up to the Hall he found Mr Crabtree waving a letter.

"I've had some good news, Sam!" he cried. "My grandchildren are coming for a visit!"

He explained how he had seven grandchildren, who lived on the other side of the world, in Australia. They had come to Britain a few weeks earlier and were coming to see their grandad on Saturday.

"I want it to be a very special day," he said. "Mrs Crabtree is going to make her four-layer chocolate cake . . ."

It was then that Sam had his brilliant idea.

"MR CRABTREE! Why don't we try and get the train going? It would be so exciting for them!"

"It would give 'em a day to remember all right!" said the old man slowly. "Do you think we could do it, Sam?"

Chapter Four

The first thing to do was to get some coal. Mr Crabtree had some at his cottage so Sam collected a bucketful. Getting the fire going in the engine was not as easy as it looked. It took Sam and the old gardener several tries to light the fire. To their delight, however, the train was soon puffing out small cotton-wool puffs of smoke, just like she used to. With a "Toot TOOT!" of the whistle, the little red train was off. The old gardener stood watching as engine-driver Sam went tearing past.

Sam made great plans for the grandchildren's visit. He found an old peaked cap in the garage which he could wear. He made tickets to 'sell' to the passengers. Aunty Kathleen made a green flag for Mr Crabtree to wave, and Uncle Dennis bought a whistle for him to blow.

"I think we're all ready for them now," said Sam on Friday evening, giving the engine a final polish.

"Red sky at night, shepherds' delight," said Mr Crabtree, looking

at the sky. "Should be a fine day
tomorrow, Sam!"

Saturday dawned a perfect summer's day. Sam hurried up to Barton Hall to be there when the children arrived.

When the children saw the little train they were so excited. They queued up at the 'ticket-office' (the garden shed) and bought tickets. Different parts of the garden were stations; there was Rose Bed City, Fountainville, and Compost-Heapton.

Mrs Crabtree served the four-layer chocolate cake, then the children all went back to the train.

"GRANDAD!" they all cried. "This little train is BONZA! We've got nothing like this at home!"

Mr Crabtree beamed in delight.

Everyone was so busy enjoying
themselves that no one noticed the
tall figure of a man standing in the
shadow of a cedar tree. It was Mr
Crossman, the owner of Barton
Hall.

Mrs Crabtree was the first to spot him. A silence fell as he came across the lawn.

"Oh, Mr Crossman!" she cried, all in a fluster. "We weren't expecting you for another week! These are my grandchildren – you did say it would be all right for them to come for a visit?"

"Yes, yes, Mrs Crabtree, that's quite all right. I'm very glad to see you all enjoying yourselves . . ."

Mr Crossman went back indoors. But for a long time he stood at a window, watching the children play.

Sam was very quiet next morning at breakfast. He was so thoughtful Uncle Dennis said, "Penny for your thoughts!"

Sam was thinking that now Mr Crossman was back at the Hall, he wouldn't be able to go there any more. He wouldn't see the little steam train ever again.

Chapter Five

Sam spent the next few weeks
swimming and taking trips around
the Devon countryside with his aunt
and uncle. Suddenly it was the last
day of the summer holidays. Sam
took his bucket and spade and went
down to the sea. The tide was
coming in and there was only a
narrow strip of beach left. He sat
throwing pebbles into a rockpool.

Just then Sam heard something
which made him jump up as if he'd
had an electric shock.

"Toot toot," it went. "Toot
TOOT!"

He raced across the beach,
splashing through the waves at the
edge of the sea – *it was the little train!*
It was puffing along the sea-front
just like it used to!

46

Sam stood watching, unable to believe it.

"Grand sight, isn't it Sam!" said a voice. It was Mr Crabtree, standing beside him.

It seemed that when Mr Crossman had seen them all that Saturday, using the train properly for lots of children, it had made him realise that it was selfish to have it just for his own family. The little train should be for *all* children to enjoy . . .

"So you see, Sam," said the old man, "It's really thanks to you that Ted's Train is running again!"

They watched as the little red steam-engine came hissing into the platform, letting out its long "OOOUUUFFF!" of steam as it came to a stop.

Ted gave Sam a cheery wave.

"Jump on, mate!" he shouted. "I hear you're a real engine-driver now!"

There was a smudge of oil on his cheek, polish on his shirt and his hands were black with coal-dust; but Sam had never been so happy. Puffing along by the edge of the sea, the wind in your hair — this is the life, thought Sam!

NO CAMPING FOR SAM

For my husband, Deh-ta

Chapter One

Sam was curled up by the living room fire, reading an adventure story. It was all about a boy called Tracker Tim who lived in the woods in a tent and made camp fires. Sam put down the book with a happy sigh.

"I wish we had a garden!" called Sam to his aunt who was in

the kitchen. "Then I could put up a tent and go camping. I could be like Tracker Tim!"

Sam lived in a big city in a block of flats with his Aunty Kathleen and Uncle Dennis. And there was no garden at the flats, only a car park and a row of dustbins.

"Well, now you are eight you could join the cubs," said his uncle, coming in with a tray of tea things. "The cubs often go camping."

While they munched their walnut cake Uncle Dennis told Sam about the cubs. He explained that it was a sort of club for boys.

"You learn interesting things

and go on outings," he said.

"And go camping?" asked Sam, his eyes shining.

"Yes!" chuckled Uncle Dennis. "And go camping!"

Aunty Kathleen found out where the nearest cub pack met. It was in St Mary's Church Hall, quite near Sam's home.

Sam began to go to the cubs every Friday. In charge of them was a small, jolly man with ginger hair called Mr Edwards. The cubs all called him Akela. This is the special name given to a cub leader.

The cubs played football and cricket. They learnt how to tie knots and use a compass. Sam loved Friday evenings!

Then came the great day when Sam was to be invested. This meant he had to make a promise to Akela to be a good cub. Uncle Dennis had bought a brand new uniform for Sam to wear, and Aunty Kathleen had sewn on the badges.

"I'm a real cub now!" said Sam
proudly, putting on the green cap
and jersey.

That Friday Sam came home
with a smart little notebook.

"Akela says that now I'm a cub
I must do a good deed every
day," said Sam. "I have to write
them down in this book."

Sam's first good deed was when
he took his aunt's washing to the
launderette. He struggled to and
fro with the big bundle of
washing.

"That's very kind, Sam!" she
said, beaming, when Sam

returned. But his aunt's smile faded away when she opened the laundry bag. "SAM! These aren't our sheets! These are full of holes!" she cried.

Poor Sam! He had brought someone else's washing home by mistake!

The next good deed was when Sam polished the wooden floor of their little hall. He got down on his hands and knees and polished very hard, much harder than Aunty Kathleen ever did. He polished until he could see his face shining back at him.

But when Uncle Dennis came home he didn't realise how slippery the hall was now. He

went skidding along on the mat
and fell, hitting his knee hard.

"Sam's been polishing that floor
all afternoon," said Aunty
Kathleen.

"So I see," groaned Uncle
Dennis.

The next good deed was when
Sam offered to clean the bath. The
only trouble was he used lavatory

cleaner by mistake. When Sam
had finished, the bath looked
terrible, all streaked with grey.
Uncle Dennis had to buy some
special bath paint to repair the
damage.

"Do you have to do a good
deed *every* day, Sam dear?" sighed
his aunt. "Couldn't it be just every
week?"

"How about once a month?"
said his uncle, rubbing his sore
knee.

Chapter Two

Sam ran all the way home one
Friday after cubs, clutching a
letter. He was too impatient to
wait for the lift so he ran up the
stairs to the fifth floor. He sat
down in the kitchen, all puffed
out.

"What does it say?" said Sam,
giving Aunty Kathleen the letter.

It was all in joined-up writing and
Sam wasn't very good at reading
that yet.

His aunt put on her glasses.

"It's from Akela," she said. "It's
about going camping with the
cubs!"

She read the letter out loud.
The St Mary's Church cubs were
going camping in Wales.

"I thought people camped in

tents, not whales," joked Uncle Dennis.

They all laughed.

"Wales is beautiful," said Aunty Kathleen, putting down the letter. "I used to go there for holidays when I was a child. There's mountains and woods . . ."

Sam's eyes lit up: to go camping in tents in real countryside! It was just what he'd always wanted. He

Needed for camp:
Rucksack
Sleeping bag
Plate and Mug
2 Towels
First Aid Kit
Boots
Torch & Battery
Swimming suit
Anorak
2 pullovers
Washing bag
Knife and fork
Compass
Soap & flannel
Pants & vests

could be like the boy in the
adventure story, Tracker Tim!

"Can I go?" pleaded Sam.

His aunt picked up the letter
again. The holiday would cost
quite a bit of money. And there
was a long list of all the things

63

Sam would need, like a sleeping bag and a rucksack. Aunty Kathleen looked worried. She didn't have much spare money these days, not since she had lost her job as a dinner lady.

But Uncle Dennis said *of course* Sam must go! They'd manage the money somehow. They talked about it over a tea of fish and chips.

"I think Bill Bateman has a sleeping bag Sam could borrow," said Uncle Dennis. Bill lived in a ground floor flat in their block. He was Sam's special friend.

"I'll go and see Bill tomorrow," said Sam, his mouth full of chips.

When they had all had their

second cups of tea Sam suddenly
jumped up.

"Aunty! I haven't done my good
deed today! Shall I wash up?"

His aunt and uncle sat watching
the television in the living room

while Sam washed up. Uncle Dennis turned up the volume on the TV.

"That way we won't hear the sound of breaking china," he explained to Aunty Kathleen.

Chapter Three

Sam went to see Bill Bateman the
next morning. He told him how he
was going camping with the cubs
in a few weeks' time and would
need a sleeping bag. Did Bill have
one he could borrow? It turned
out that not only did Bill have a
sleeping bag, but a fine rucksack
as well!

Sam tried on the rucksack for size.

"It's almost as big as you, Sam!" laughed Bill.

It was a super rucksack, with two big side pockets and a little Union Jack flag sewn on the top.

"Thanks, Bill! It's great!" said
Sam, grinning from ear to ear.

Bill's wife Doris gave Sam a
little tin mug.

"You can paint your name on
the side, Sam," she said. "I don't
think Bill will be needing his

camping things any more . . ."

Sam began to get together all the other things he needed for his camping holiday. His aunt ticked them off on Akela's list. Uncle Dennis bought him a fine torch which was Sam's pride and joy. He kept turning it on and off.

"You'll use up all the battery before you get to Wales!" warned his uncle. Sam put the torch away with the sleeping bag. He studied Akela's list for the hundredth time.

"I still need the first-aid kit, Uncle," said Sam.

At last everything was ready. The
big rucksack was packed and sat
in the corner of the hall, almost
filling it up. The torch was in one
pocket, the first-aid kit in another.
The little tin mug dangled by a
string, with his name, *SAM*,
painted in red letters on the side.

That night Sam was so excited
he couldn't get to sleep. He kept
thinking how this time tomorrow
he would be like Tracker Tim. He
would be camping in a tent!

It was almost midnight when
Uncle Dennis put his head round
the door of Sam's room.

"Not asleep yet, old chap?" he
said.

Sam tossed and turned in his bed.

"I don't feel very well, Uncle Den," said Sam. "I feel so hot and my skin is all itchy!"

Uncle Dennis put on the light. Sam's eyes were strangely bright and his cheeks were very red. And all over his skin were hundreds of tiny pink spots!

Uncle Dennis rushed to fetch Aunty Kathleen.

She took one look at Sam and knew at once what was the matter: MEASLES!

"No camping for Sam, I'm afraid," Sam heard Uncle Dennis say to Aunty Kathleen, as he went out to fetch a cool glass of lemonade for the patient.

Poor Sam! He felt like crying.

Chapter Four

For a week Sam was very ill with
the measles. Dr Spencer came
three times and gave him pink
medicine. Uncle Dennis read him
Rupert Bear books and Aunty
Kathleen cooked his favourite
food. Everyone was very kind.
Only Sam wasn't a bit hungry

and he couldn't enjoy the Rupert
stories, he felt too ill.

Then, after a week, Sam began
to get up and about again. But all
he did was sit around and mope,
feeling cross and spotty. And he

couldn't help thinking of the cubs,
on their camping trip in Wales . . .
Sam didn't even have the
energy to do a good deed every

day now. His aunt and uncle began to get quite worried.

It was late one afternoon that Uncle Dennis came home with a mysterious bundle under his arm. He told Sam not to come into the living room until he called him. He had a nice surprise, he said, to cheer Sam up!

Sam heard bumps and giggles as his aunt and uncle arranged his nice surprise. And when they called him to come in, Sam couldn't believe his eyes — they had put up a dark green tent in the living room! One guy rope was tied to Uncle Dennis' heavy chair, and the other to the sideboard!

"Where did you get it?" said Sam in amazement.

"Bill, of course! It's his old army tent," beamed Uncle Dennis.

Sam ran to get his torch and sleeping bag. He crawled into the tent. His uncle's round, smiling face appeared in the tent door.

"What do you think, Sam? You can sleep the night there! Then you'll be camping after all, eh?"

Uncle Dennis went off, chuckling to himself.

Sam sat there, huddled in the sleeping bag, all alone in Bill's old

tent. How kind his aunt and uncle were! Trying to cheer him up like this! But no matter how hard he pretended, a tent in the living room just wasn't like proper camping. It just wasn't the same, thought Sam. He longed for the countryside, for woods and streams and a real camp fire. Sam found that a tear was rolling down his cheek, and he had to blink back more.

Just then there was a knock at the front door and Sam heard Bill come in. He knew it was Bill by the squeak the wheelchair made on the wooden floor of the hall.

There was someone with him, too. Someone with a high, shrill

voice. Bill had brought his nephew
Toby with him, a little boy of just
six.

"Toby would love to see the
tent up," said Bill's voice.

Sam heard the living room door
open and Toby crawled into the
tent.

"Brill!" breathed Toby. "Aren't
you *lucky*! Staying the night in
Uncle Bill's army tent!"

Sam showed Toby the torch,
the compass, the first-aid kit and
all his camping things. Toby was
delighted with everything. He
especially liked the knife, fork and

spoon set, that slotted together into a little case.

"Aren't you LUCKY!" Toby kept saying enviously.

Then Aunty Kathleen asked if Sam would like his tea in the tent? And would Toby like some too? Toby's eyes shone!

After they had eaten their baked beans on toast Sam told Toby all about Tracker Tim and his adventures.

"Let's play that!" cried Toby excitedly. "We'll pretend we're in the jungle. You can be Tracker Tim, and I'm the tiger . . ."

They had a great game. Then Sam showed Toby some of the things he had learnt at cubs about

camping. He showed Toby how to
find North on the compass (it was
in the direction of the sideboard).
He showed him how to tie a
special knot in some string.

Then Sam spotted his aunt's
vase of flowers and twigs. It gave
him an idea. He told Toby he
would show him how to do
'tracking'. He explained how cubs

leave trails by making signs, laying twigs in the shapes of arrows.

"Tell you what! I'll lay a trail round the flat which will lead you to water!" said Sam.

Soon Toby was following a trail of twig arrows till he came to the bathroom.

"It led me to water!" he cried triumphantly.

Chapter Five

That night Toby was allowed to
stay in the tent with Sam (luckily
he had already had the measles).
Both boys slept soundly and in the
morning they ate an enormous
breakfast. They had cornflakes,
followed by sausages, then toast
and marmalade.

90

Then Bill arrived to take Toby home.

"I think you'd better keep that tent, Sam," said Bill. "I have a feeling you may be going proper camping one of these days!"

Toby poked his head out of the tent.

"Didn't you know, Uncle Bill? Sam and I *did* go proper camping! ALL NIGHT!" said Toby indignantly.

Sam smiled to himself. Funny Toby! Sam wondered if *he* had been like Toby when *he* was six years old . . .

But Toby had really cheered Sam up: after all it had been fun, camping in the sitting room and

pretending to be Tracker Tim!

Bill showed Sam the correct
way to take down the tent.

"Then next time you go

camping, you'll know how to do
it," he said.

"Yes," smiled Sam, "I'll know
just how to do it!"

When Toby and Bill had gone
home, Uncle Dennis washed up
the breakfast things while Aunty
Kathleen dried.

"Sam seems to have got his
appetite back, anyhow!" said his
aunt. "He had five sausages!"

"Yes!" beamed Uncle Dennis.
"My idea worked, didn't it? You'll
see, Sam will soon be back to his
old self again now!"

Just then a low rumbling noise
came from the living room. Uncle
Dennis and Aunty Kathleen
rushed to see what had happened.

Sam was sitting on the floor
amid a pile of books.

"I was only dusting, Aunty —
doing my good deed for the day
— and the books just fell on top of
me!" he grinned up at them.

His aunt and uncle looked at each other and burst out laughing. "Yes!" they both said together. "Sam's definitely back to normal!"